"The Sky is Falling!"

A Global Warming Survival Guide

by

Cal Orey and Mark Jabo

illustrated by
Kelley Cunningham

Bloomington, IN  Milton Keynes, UK

authorHOUSE®

AuthorHouse™
1663 Liberty Drive, Suite 200
Bloomington, IN 47403
www.authorhouse.com
Phone: 1-800-839-8640

AuthorHouse™ UK Ltd.
500 Avebury Boulevard
Central Milton Keynes, MK9 2BE
www.authorhouse.co.uk
Phone: 08001974150

First published by AuthorHouse 2/16/2007

ISBN: 1-4259-6943-7 (sc)

Printed in the United States of America
Bloomington, Indiana

This book is printed on acid-free paper.

Contents

Acknowledgments

I dedicate this book to two intelligent mammals–
humpback and blue
whales–who teach us that man is not invincible nor incapable of
ending the planet.
The motto: We all can learn from Moby Dick, Humphrey and
Willy–the last great American whales.
–Cal Orey

I dedicate this book to man, the rational animal, who only gets in
trouble when he doesn't take the time to think for himself. Or, as
my grandfather used to say, "You can go far in life...you just need
to learn to separate the fly crap from the pepper."
–Mark Jabo

Foreword
Global Warning

When we look back from year to year
We lived our lives without much fear.

The streams were kind to wade and swim;
The trees would beckon with each limb.

The coastal terrace never failed;
The beaches all remained "unwhaled."

The hurricanes stayed out to sea;
Tornadoes never made us flee.

And earthquakes rarely topped a "four";
For homicides, no one kept score.

But now the glaciers melt too fast;
The deserts all become more vast;

The oceans rise, the levees fail;
Instead of showers, we get hail.

As frictions rise on every land,
We wonder what the Lord has planned.

With Global Warnings everywhere
It's time for people to prepare.

by

Jim Berkland
Geologist
www.syzygyjob.com

PART ONE

Ch-Ch-Ch-Changes

CHAPTER 1

I Feel the Earth Move

"If you asked me to name the three scariest threats facing the human race, I would give the same answer that most people would: nuclear war, global warming and Windows."
–Dave Barry

Chances are, you're already aware of the signs documented in *Time Magazine*:

"In Africa, drought continues for the sixth consecutive year, adding terribly to the toll of famine victim... record rains in parts of the U.S., Pakistan and Japan caused some of the worst flooding in centuries.... A series of unusually cold winters has gripped the American Far West, while New England and northern Europe have recently experienced the mildest winters within anyone's recollection.

As they review the bizarre and unpredictable weather pattern of the past several years, a growing number of scientists are beginning to suspect that many seemingly contradictory meteorological fluctuations are actually part of a global climatic upheaval."

The question remains.... What does the changing climate mean to you and our planet?

When the Sky Comes Falling Down

Is the sky *really* falling? If you haven't heard by now, listen up. The end of the world is near. Any idiot can see that, right? *Time Magazine* is only one of countless publications to tell you what everyone already knows: there's a crisis of global proportions and it affects us all.

The coming climatic cataclysm is the new boogeyman on the block and is so serious it's going to get you, your children, and your children's children. It will also affect your second cousin, your cousin's cousin, your children's second cousins, your children's second cousins from a previous marriage, your hamster's hamster, and the clone of your hamster's clone. Whew. Did we leave anyone out?

They say that those who don't know history are doomed to repeat it. Of course, they said the same thing about eighth grade algebra, but somehow most us still managed to pass the course (despite the lack of a natural disaster or nonexistent alien to abduct the classroom teacher) and graduate.

Chances are, we'll probably weather the latest global climatic Armageddon, too. Because you see, the apocalyptic quote we started with *is* from *Time Magazine*.

But, as it turns out, it's from the June 24, 1974 issue when the concern was, if we didn't do something immediately, we would be in another Ice Age within 30 years. Back then it was crucial that we act because if we didn't, there would be global disasters of unimaginable proportions. You know, real wrath-of-god type stuff.

Only one problem. The fear mongers and doomsayers were wrong. This should come as no surprise to anyone who has ever left the house with an umbrella, only to carry it around all day without using it. As they have for centuries, weather forecasters continue to get it wrong.

If you're human, you might not think about global warming 24/7. Here, take a quick peek at some hot terms that we decoded for you.

The Heat is On Lexicon

Hot Terms	Definition	Rose-Colored View	Doomsday Scenario
Global Warming	The increase in the Earth's average temperature over time	Endless summers, dude	We're toast
Greenhouse Effect	The way the Earth's atmosphere warms the planet	What's all the fuss? All kinds of things grow in green-houses	Ever burn popcorn in a microwave? That's us about five more years
Fossil fuels	Coal, oil and natural gas	What the Flintstones used	Part of the plot by the oil companies to control the world
Greenhouse gases	Gases in the air that cause the greenhouse effect	Just a lot of hot air about climate change	What those evil corporations emit
Kyoto Protocol	International treaty for the mandatory reduction of greenhouse gases	When you take your shoes off before entering a Japanese home	Only a few billion dollars for a good cause

Next, take an up-close and personal peek at the history of optimistic global warming "Pollyannas" and pessimistic "Chicken Littles" who have squawked "The sky is falling!"–and like the beggars in *Waiting for Godot*–waited and waited for something to happen....

The History of the World

When talking about global climate change in a nutshell, there's a lot of history to chat it up about. There have been 17 ice-age cycles in the past two million years. To put this number in perspective: that's more than four times the number of *Star Trek* movies–including *Star Trek IV: The Voyage Home*–in which, coincidentally, James T. Kirk and crew save mankind once they discover the strange sound is the call of the humpback whale. Our two whales, Harmony and Hamlet, will help serve as guides through the maze of global warming hype and hysteria. Scientists believe that whales (intelligent, warm-blooded mammals that they are) often get disoriented and beach themselves during erratic earth changes–such as when our planet is getting hotter.

Whale behavior aside, not much has changed over the past two million years. People are still talking about the weather. Everyone from politicians to some of today's biggest Hollywood celebs are concerned with what is being described as a looming global crisis. Critics of the *Star Trek* franchise aside, many of these people are concerned about global warming, too.

Some scientists and the majority of reality TV show contestants want to believe that an Ice Age wiped out the dinosaurs in an early version of *Survivor: Jurassic Era*. This raises the question, "Could an abrupt climate change threaten human life?" Most dinosaurs were large creatures with brains roughly the size of a peanut. It is unlikely that humankind, with our technology and opposable thumbs, would fall victim to the same fate, although you could argue that fans of professional wrestling fit the "large-creatures-with-brains-the-size-of-a-walnut" profile that proved so unlucky for Stegosaurus and friends. Unlike dinosaurs, whales have the largest brain of any animal, which is why we chose them as our spokespeople, er...spokeswhales.

When we talk about Ice Ages, do you think about a time when extensive ice sheets covered large parts of North America ore Europe?

The most recent Ice Age ended 10,000 years ago as the Pleistocene Epoch drew to a close, drawing rave reviews and ending a streak that would stand until it was topped by the 12 Tony award-winning Broadway run of *The Producers*.

Blast from the Past

Today, in the 21st century, we are in the Holocene Epoch (from the Greek "holo" meaning "hollow" and "cene" meaning "scene") which many paleontologists believe helps to explain the recent proliferation of reality TV shows. The real truth is, you don't have to turn on the weather channel because you're living it.

More bad news: every epoch (or in technical terms, "really, really, long time") has experienced glacial advances and retreats. Glaciers were originally thought to operate in 40,000-year cycles, but with advances in modern science, are now believed to advance and recede in 100,000-year cycles. Since the last glacial cycle ended 10,000 years ago (give or take a few days), there's a chance we can look forward to a gradual shrinking of the world's glaciers for the next 90,000 years. The worst part about this? It's probably inevitable there will be *An Inconvenient Truth, Part Two*.

Indeed, history can and does repeat itself. In more recent times, global temperatures have continued to fluctuate. Proxy data tell us there was a Medieval Warm Period that lasted from 800 A.D. to 1300 A.D. Since there are no accurate records, scientists use proxy data to estimate temperatures going back over 1,000 years. Proxy data include such things as tree rings, fossil records and first-person accounts from Joan Rivers.

After the Medieval Warm Period, the global climate cooled by nearly two degrees centigrade during the period from 1400 to 1900. Climatologists refer to this period as the Little Ice Age. The Little Ice Age strongly objects to this term and has repeatedly maintained it's not the size of your glaciers that matters; it's how you use them.

Baby, It's Cold Outside

Since the end of the Little Ice Age, temperatures have been getting gradually warmer throughout the 20th century up to the current day.

Well, except for a period between 1940 and 1975 when temperatures got cooler.

When the first Earth Day was held in 1970, it was global *cooling* that was all the rage. The consensus was another Ice Age was imminent. Some scientists, using carefully constructed climate models, predicted that glaciers would cover most of the U.S. by the time the year 2000 arrived. (Y2K was also supposed to end the world as we knew it.) These were scientists, and they had models. Back then, *Time* and any national magazine worth whale meat (more expensive than ever due to diminishing supply) would have told you there was a consensus among all these right-thinking scientists that we would see another Ice Age before the next century.

Remember how cold it was back in the 1990s? Neither do we. Because there were no giant glacial advances, no wooly mammoths, and, if you could dress yourself, you were probably pretty comfortable, no matter if you lived in the southern or northern hemisphere.

Somehow, without missing a beat, scientists changed their position in mid-stream. Like disoriented whales who can't deal with earth changes and simply beach themselves, scientists were suddenly predicting global *warming*. But wait, here's the best part...global warming and global freezing supposedly were both caused by the same thing: too much carbon dioxide. Where did all this excess carbon dioxide come from? It came from airplanes, factories, SUVs and humans burning oil for energy. Or, to put it simply, progress.

The glitch is humans and animals need carbon dioxide to live. If there's not enough, things get too cold. If there's too much, we'll be too hot. And like Goldilocks and the Three Bears, with humans burning all these fossil fuels, it's very hard to get it juuuussst right. The concern of the moment: our planet is getting warmer and warmer. The end result: cooked humans and whales.

Now that you've had a crash course in Climate 101, let's take a look at some of the details of the how hot the earth might get and how it will affect your life today and tomorrow.

Last Great American Whales
Speak Out

All is not well.[1]

Hamlet Says: I admit it. As a gray humpback whale, I am afraid. I am really afraid. I can stand to lose a few pounds due to all the krill (shrimp-like crustaceans) I devoured in the good old days. But now, the word is, there's a big shortage of whale food on the planet. I don't believe it's due to the world getting hotter.

Meanwhile, while two-leggers munch on processed food, drive their gas hogs, and develop more ivory towers, my hungry whale pals are becoming an endangered species. According to people who study the earth, a whale food shortage is just the beginning. It's a mega global warning for doomsday. Read: drought, earthquakes, flooding, hurricanes, landslides, severe weather changes, solar flares, tsunamis, volcanoes, wildfires, plus melting ice wreaking havoc on our lives as we once knew it. So, the question is, are we to be dead whales on the rocks or not to be? I don't want to think about it. It can't be true.

Harmony Says: There are too many melodramatic global warming alarmist whales in the sea. The fact is, ocean tides ebb and flow. Personally, as a creative and cerebral blue whale, I don't like to believe humans are the culprit of climate changes. It's part of our planet's life cycle. Scientists say floating ice shelves which surround much of Antarctica–our fave spot–were happening decades ago.

I hear a lot of humans talking about a collapse of Antarctic ice shelves leading to a rise in worldwide seas levels. I like to look on the bright side: that means, we whales will have our krill and eat it too–and have more open water to swim in. Global warming, global schwarming. I saw a couple of skinny whales, but I am okay. It isn't that bad in Hawaiian waters during the winter and Alaskan seas in the summer, right?

CHAPTER 2

Mother Nature

"At any moment there are 1500 electrical storms on the planet...
Violent, disruptive, chaotic activity is a constant feature of our
globe. Is this the end of the world?
No: This is the world."
–Michael Crichton

Stairway to Heaven

So what *exactly* is global warming? You might think that would be an easy question for everyone to answer and agree on, right? Not so fast, young Skywalker.

When it comes to global warming, even people studying the subject have a hard time agreeing what it is we're trying to measure and what the best way to measure it is. The end result is that most scientific discussions about global warming quickly degenerate into a volley of statistics being hurled back and forth between scientists.

Listening to scientists debating statistics and statistical methods is a lot like listening to two cooks argue in a foreign language. You're pretty sure there's something going on; you're just not sure what it is and how it affects you.

It would be so much easier if scientists could just build a big thermometer and stick it someplace in the Earth for 30 seconds to get the average temperature. But, as you can plainly see, there are a number of problems with this solution. Where would we stick the thermometer–in the Arctic or in the Equator? Would we take the Earth's temperature in the summer or the winter? And, perhaps most important, who would pay the deductible if Mother Earth had to go out-of-network?

Scientists attempted to solve this problem by looking at a whole bunch of little thermometers all over the world and then taking an average of the readings. This was working very nicely until someone noticed that cities generate more heat than areas that are not as heavily developed. Known as the urban heat island effect, this quirk in the data meant that the average temperature of a given thermometer near an urban area was skewed upward over time as the city expanded.

Faced with this problem, scientists did what most of us do when our wife or girlfriend asks us to guess how much she weighs: they fudged the numbers and made all kinds of excuses as to why the temperature wasn't *exactly* what the thermometer said it was.

One Hot Minute

The same holds true when we try to figure out if the Earth is getting hotter and how hot will it get. At the end of the day, this means no one can agree exactly how much the Earth has warmed over the past 100 years or how much weight your mate has put on since you've first started dating him or her.

In both instances, we can agree that the numbers are higher. But what about the causes? Here's where all the yelling and screaming starts. Your significant other may get pretty upset, too.

There are two main schools of thought when it comes to the causes of global warming. The first school suggests that warming temperatures are the result of natural cyclical forces and human activity is only a small part in the overall scheme of things. This is the way the world has worked for the past few million years.

The second school of thought argues that humans are destroying the balance of nature and are the culprits in the Earth's temperature rise over the last 100 years. Evil inventions like airplanes, electricity and the lawnmower are creating a "greenhouse effect" that will ultimately cook the planet to a crisp.

So, which school of thought do you think would get more play in the press?

Has the average temperature of the planet risen over the past century? Yes, it has…by a whopping whole half a degree centigrade. Interestingly, nearly three-quarters of that rise took place prior to the 1940s–well before any major man-made emissions of fossil fuel. That fact doesn't exactly jibe with the industrial-man-as-the-cause-of-global-warming theory but, no matter; throw in some news footage of natural disasters and a deep, authoritative voice-over blaming it all on man's violation of Gaia, and people will forget about actually checking to see if it all makes sense.

Are we saying that humans haven't contributed at all to the global rise in temperatures? Of course not; we probably have. But so have Golden Retrievers–there are more of them than ever and they throw off a lot of heat after a good run in the park. Oh, and by the way, trees, clouds, sun-spots, cyclical ocean currents and a ton of other things we can't even begin to understand have also contributed. However, it's just not as sexy as blaming big corporations, greedy politicians and SUV-driving junior executives.

It's Too Late

But what about the future? Some folks say it's too late to stop global warming–others believe we can delay the end of the world as we know it. How much hotter is it going to get over our lifetime? The best estimates scientists can come up with range from two degrees to 15 degrees over the next 100 years.

If you went to a fortune-teller and she told you you'd have from two to 15 kids or get married two to 15 times during your life, you'd walk out of her shop laughing about what a crock the whole crystal ball industry was. Of course, you wouldn't be chuckling later that evening when you grew a tail, but that's another story.

Suffice to say, two to 15 degrees is an incredibly wide range of guesstimates about how much global temperatures could change over the next five or six decades. (If it makes you feel any better, the most credible estimates are clustered at the low end of the range.)

"We have about five more years at the outside to do something."

Although that sounds like a quote from an HBO special about global warming, it's actually from a 1970 speech at Swarthmore College by ecologist Kenneth Watt. Mr. Watt went on to say, "The world has been chilling sharply for about 20 years…. If present trends continue, the world will be about four degrees colder for the global mean temperature in 1990, but 11 degrees colder in the year 2000. This is about twice what it would take to put us into an ice age."

Remember the Great Ice Age that occurred at the turn of the century? Neither do we. Somehow, despite warnings that "a new ice age must now stand alongside nuclear war" as a risk to the planet, we somehow made it through the 1970s, '80s, '90s all the way to the present. Turns out cable TV monopolies were a greater threat than climate change.

Anticipation

If it's not too late to turn around this doomsday scenario, just how long do we have, anyhow? Nearly 40 years later we're still talking about climate change (to use the politically correct term). Only this time, it's global *warming* that's got everyone's lab coat in a twist. Both opponents *and* advocates of the Kyoto Protocol agree that the impact of total implementation of the treaty is the *potential* to knock somewhere between .02 to .07 degrees centigrade off the predicted global warming at a cost of *trillions* of dollars. For $100 trillion, we could save a whole degree. Or you could just stop by your neighbor's pool.

Politicians and scientists are talking about putting gigantic mirrors into space to reflect away some of the sun's heat in an effort to minimize global warming. This ignores the major problem with trying to control climate change: who's going to be in charge of the thermostat?

Think about it. The six people you work with can't agree on what temperature to keep the office. What kind of problems are we going to have when the whole world is involved? You know that while we're sleeping, North Korea will be up in the middle of the night turning the thermostat down just to mess with us.

"Climate change is real, and it affects all of us," the sonorous voice on the news warns us. Of course, it's real, and, of course, it's changing. That's what climate does and has been doing for billions of years. The folly is we're still trying to outguess and control the weather. So, in that respect, humans really are the cause of the problem.

Meanwhile, as you decide to be a global warming skeptic or climate crusader, the weather report continues and life goes on....

Last Great American Whales
Speak Out

Here comes the sun, ♪ here comes the sun, and I say it's alright.²

Hamlet Says: *Whale News, WNN* and *Whalesday* reporters say global warming may affect both humans and animals. Heat stress and other heat-related woes are caused by warm air and high humidity. In many places (maybe even the Garden Island), global warming will increase the number of hot days during the year. But maybe these people are making it sound big to sell magazines, newspapers, and advertising.

As the rumor goes, heat rises, air pollution, changes in food and water supplies, and coastal flooding might disorient creatures, big and small. That means our young calves and senior whales will run the highest risk. What if we whales have to migrate to different locations? What if the sun gets hotter? What if it's all a hoax like other end-of-the-world-scenarios? Or what if we really are close to "The End"?

Harmony Says: I've heard from geologists that landlubbers aren't to blame for global warming. Actually, it may be simply part of the Earth's cycle like the four seasons. Sure, mankind helps the temperature rise. But, is it really that bad? As a whale with roots that go back 50 million years, I know for a fact that the layer that insulates the surface of the Earth from the Sun's rays helps to regulate temperatures on the planet.

So these days, we are in an "interglacial period" that began 10,000 years ago. In other words, we whales and humans (if they don't exterminate themselves first) may or may not witness an Ice Age as the cycle goes around. But, rather than point the finger at the sun, what about other big threats such as entanglement in fishing nets and illegal whaling? Global warming is just one more worry to put on the back burner and ignore. Maybe it will go away.

PART TWO

What's Going On?

CHAPTER 3

Daydream Believer

"A consensus means that everyone agrees to say collectively what no one believes individually."
– Abba Eban

A lot of people believe there is hard-hitting scientific consensus regarding global warming. Then again, a lot of people believe in Scientology, so go figure. Decide for yourself. We brought together two of the most popular cults for a showdown to see which would become the new religion of choice. It's Scientology vs. Global Warming. Feel free to keep score at home.

Category	Scientology	Global Warming	Round Goes to:
Seminal document	*Dianetics*	*United Nations IPCC Report*	**Scientology** Quality of science is similar Recognizable cover art favors Dianetics
Propaganda movie	*Battlefield Earth*	*The Day After Tomorrow*	**Scientology** Claims made in *Battleship Earth* are more believable
Money comes from	Gullible people	Weak-willed corporations	**Global warming** CEOs should know better
Creepy spokesperson	L. Ron Hubbard	Al Gore	**Tie** Both have padded résumés
Feel good mission	Rehabilitation of the human spirit	Sustainable development	**Global warming** Scientologists couldn't convince you to sort your own garbage
Technology	Stress meter	Hybrid car	**Global warming** Hybrid car owners only slightly less obnoxious than Scientology recruiters

Category	Scientology	Global Warming	Round Goes to:
Celebrity shills	John Travolta, Kirstie Alley, Tom Cruise, Catherine Bell, Leah Remini	George Clooney, Julia Roberts, Cameron Diaz	***Global warming*** Compare box office receipts
Movie line that could be describing supporters	"She's stupid enough not to be a menace, good-looking enough to be decorative and gets drunk with economical speed.... -*Battlefield Earth*	"What about the garbage? Always something to eat in the garbage!" -*Day After Tomorrow*	***Scientology*** It should be obvious
Lie repeated until it's taken as truth	10 million worldwide followers	There's an overwhelming scientific consensus regarding global warming	***Global warming*** More widely believed despite being easier to disprove

Trust in Me

So, as the chart shows, global warming stacks up as a contender in the contest to be the religion of choice for those people who are looking for a trendy new cause.

This brings us to the topic of "scientific consensus." Scientific consensus is an oxymoron that has found its way into popular jargon, kind of like "private e-mail" or "U.N.-designated safe haven."

Science deals in facts. Or at least it does until it becomes politicized. When that happens, people start using the phrase "scientific consensus" as way to stifle debate. "Since you're not smart enough to figure this out," they tell you, "just take the word of these people that I've decided know what they're talking about." But what if they are wrong? The whole reason we have science is to prevent this type of intimidation by experts. Science is the warm comfort of knowing that something is true without having to wade through the spin. No one talks about a scientific consensus that gravity works. Or that a sperm fertilizing an egg leads to babies. There's no need to, because these things are facts.

When people invoke a consensus, they may be trying not only to cut off discussion but may also be refusing to admit how thorny the problem is. When we refuse to appreciate the complexity of something, reality generally takes its revenge through the law of unintended consequences. Here is a close-up look at history that proves it.

Plastic People

We all want to hang out with the cool group. Ever wonder what the consensus was in the past? Stop thinking. Take a look at 10 hot moments in history where people didn't think for themselves and just jumped on the bandwagon.

10. Consensus: The *Titanic* is unsinkable. Using the most advanced technology of its day, the *Titanic* was supposed to be the first in a group of super-liners that would dominate trans-Atlantic travel. **Winners:** James Cameron

Losers: Thomas Andrews, the ship's builder who was on board to assess its performance. Macauley Culkin who was rumored to be in line for the lead role in the movie that went to Leonardo DiCaprio.

Unintended consequences: One of worst maritime disasters in history propels career of Celine Dion (regarded by some as one of the worst land-based disasters in history).

9. Consensus: The world is flat. Widely held belief was responsible for stunting exploration beyond the known world.

Winners: Christopher Columbus who parlayed a trip to Spain into fame, fortune and enough 15th century groupies to last a lifetime.

Losers: Guys who have to clean up after the annual Columbus Day parade.

Unintended consequence: New World goes on to become cradle of freedom, world superpower and fattest nation in history.

8. Consensus: It's Raining Men. Song widely believed to have been responsible for more hook-ups than Time-Warner cable.

Winners: The Weather Girls and gay men everywhere.

Losers: A confused young man who had a very awkward morning-after breakfast and years of therapy.

Unintended consequence: A very awkward morning-after breakfast and years of therapy.

7. Consensus: Global Cooling. Surprise! Same stuff, different decade. Back in the day, this was the hipsters' choice for how the world would end.

Winners: Environmental activists and green politicians who were able to escape with credibility unscathed.

Losers: The Charlie Browns in all of us who are lining up to take another run at the global climate disaster football.

Unintended consequence: Al Gore mentioned as Oscar candidate.

6. Consensus: Gore wins Florida. Despite early media consensus, majority of Supreme Court justices rule Bush winner in hotly contested election.
Winners: Bill Maher, late-night talk show hosts, Fox News.
Losers: Yale.
Unintended consequence: Too numerous to mention.

5. Consensus: Beta technology superior to VHS. First in a long line of lessons for gadget freaks that marketing and distribution count just as much as technology.
Winners: X-rated film buffs who have choice of two formats. U.S. economy as people forced to upgrade to DVDs a few years later.
Losers: Movie buffs who are forced to buy DVDs a few years later.
Unintended consequence: Sex films become mainstream as Traci Lords crosses over into legitimate film work

4. Consensus: Population explosion will doom mankind. In the late '60s and early '70s, scientists and intellectuals forecast global Armageddon due to world's rapidly expanding population.
Winners: Paul Ehrlich, Stanford professor whose 1968 book, *The Population Bomb*, forecast mass starvation in America and Europe before 1980. Tom Monaghan who, in the same year, became sole owner of Domino's Pizza. By 2005, Dominos had 7,875 stores in more than 55 countries, approximately 145,000 employees, and revenues of $1.5 billion.
Losers: Stanford students…Ehrlich had tenure.
Unintended consequence: Humanity flourishing.

3. Consensus: The Earth revolves around the Sun. Nobody expected the Spanish Inquisition. And nobody expected them to be wrong about the nature of the Universe. After all, it was in the Bible.
Winners: All of us as Galileo becomes the "father of science" despite imprisonment and threats from the religious right of his time.

Losers: All those kids in school in the 1600s who had to re-do their science projects and make new Styrofoam and hanger models of the solar system.

Unintended consequence: Boy becomes a man on class field trip to the Hayden Planetarium, nearly 365 years to the day after Galileo's discovery of Jupiter. (Thanks, Betty Ann.)

2. Consensus: The Red Sox will never win a World Series. Curse of the Bambino had been raised to near dogma after over 80 years of post-season frustration.

Winners: Major league baseball.

Losers: Barry Bonds. Nothing to do with this…just because.

Unintended consequence: Boston Tea Party no longer ranks as the town's most raucous public demonstration.

1. Consensus: Global warming. Rapidly giving Orson Welles' *War of the Worlds* a run for its money as the greatest media hoax of all time. Despite repeated announcements labeling it as a work of fiction, people thought that the world was in imminent danger. And the same goes for Orson Welles' radio hoax.

Winners: Politicians who see votes and money; sensationalist media outlets that recognize a good story when they see one.

Losers: Other lifesaving research that could be funded with money we're wasting on something we can't predict or control.

Unintended consequence: Al Gore resurrected.

Okay. Now that we've put the power of mass persuasion on the table, let's discuss the role of sensationalism on the big and little screen….

Last Great American Whales
Speak Out

There is nothing either good or bad, but thinking makes it so.[3]

Hamlet Says: We can't believe everything we hear. But when journalists, scientists, films and documentaries show us the worst-case scenario–it's scarier than choking on plankton. (It's not easy performing the Heimlich maneuver on a 40-ton mammal.) Those disaster movies such as *Crack in the Ocean* and *A Whale of an Armageddon* are a whale's worst nightmare. And now, *The Day After the Weekend* is a cautionary tale for whales in Antarctica and what we have to look forward to. Ugh! It's enough to make a whale beach himself. What if tomorrow doesn't come? It's all humans' fault.

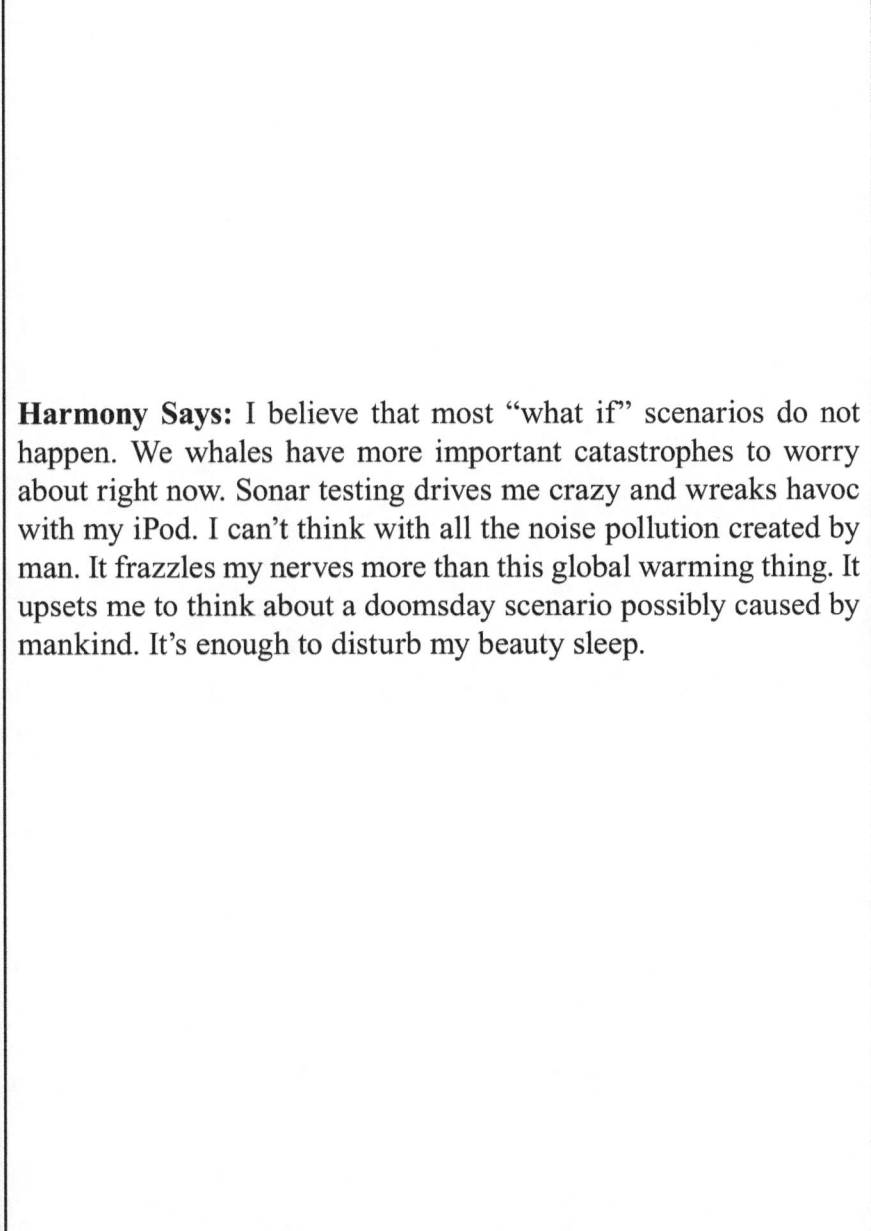

Harmony Says: I believe that most "what if" scenarios do not happen. We whales have more important catastrophes to worry about right now. Sonar testing drives me crazy and wreaks havoc with my iPod. I can't think with all the noise pollution created by man. It frazzles my nerves more than this global warming thing. It upsets me to think about a doomsday scenario possibly caused by mankind. It's enough to disturb my beauty sleep.

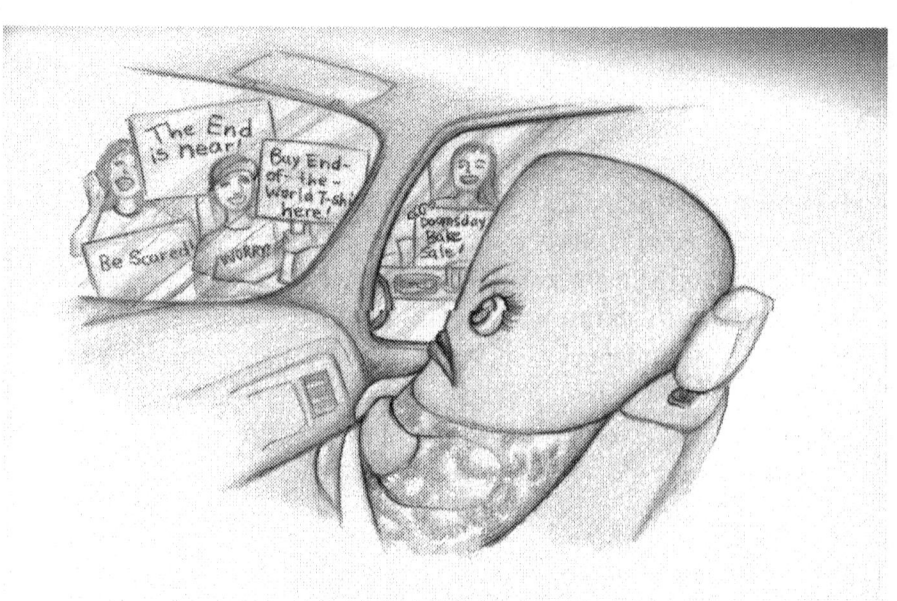

CHAPTER 4

Mother, Mother, Why Are All the People Lying?

*"The one function TV news performs very well is that when there
is no news we give it to you with the same emphasis
as if there were."*
–David Brinkley

You would think we would have learned by now not to trust
authority figures who doctor up science to come to all kinds of bogus
conclusions.

Lie, Lies, Lies
But then again, your first experience with this was probably your
mother telling you that you couldn't go swimming for at least an hour
after eating or you'd get a cramp and drown.

The eating/swimming/drowning chain of cause-and-effect was
our first brush with a scientific consensus. Everyone's mother said
the same thing. and all the mothers had the same scientific-sounding
explanation about how when you eat, blood rushes to your stomach
to help digestion.

After that, the science got a little fuzzy. Somehow, we were led to believe that if you came in contact with water, this would cause cramps and you would sink like a stone to the bottom of the river, the pond, or the shallow end of the public pool. It was a fact of nature.

The smart kids (many of whom would later go on to question the claims of climate change doomsayers) noticed a few holes in the wait-an-hour-before-going-swimming-or-you'll-die theory....

Why was it that the mysterious digestion-induced blood flow never caused cramps on dry land? What if it rained and you walked through a puddle after a big lunch? Would you immediately cramp up and drown in two inches of rainwater only to be found the next day lying in the middle of a perfectly dry sidewalk, baffling the police as to the cause of your death and inspiring a two-part episode of *CSI: Miami*?

And why was it that the one-hour rule held whether you ate a cookie or had just finished a Thanksgiving meal?

More important, why hadn't you ever heard of anyone anywhere in the world who had drowned as a result of flaunting what everyone knew to be a scientific fact? "See what happens when you don't listen to your mother?" is a phrase that dates back to the dawn of time and can be found among the early cave drawings of Cro-Magnon man. As a result, any evidence supporting the eating-drowning link would have been big news mothers around the world couldn't wait to tell to their children.

So, maybe it's not surprising that so many people are buying into the global warming hype and forecasts of climate catastrophe. It's what we've been taught to believe since we were kids. But, that doesn't mean it's correct.

Do You Believe in Magic?

Like the ESD (eating/swimming/drowning) theory, all the excitement about climate change starts with a simple fact. The ESD theory states that there is increased blood flow to the stomach as the result of consuming food. That's a fact. Likewise, earth-based temperatures have generally been rising over the past hundred or so years. That, too, is a fact.

After that, both theories veer off into wild conjecture, dubious scientific conclusions and a fair amount of pop psychology. Climate change Cassandras take the simple fact that certain earth-based temperatures readings have risen over the past century and proceed to weave that fact in with unproven theories, doctored statistics and some well-placed outright lies to construct a scenario that *sounds* scientific but has very little real evidence to back it up.

Take a look at the garden-variety global warming theory of the Earth's coming demise. As the popular formula goes: The planet is heating up…humans are burning fossil fuels…fossil fuels increase the greenhouse effect and make global warming worse. We can't afford to stand by because global warming leads to heat waves, melting glaciers, catastrophic storms, migrating viruses and the end of the world tomorrow at noon.

Okay, we made up the part about the timing of end of the world, but it really is becoming difficult to find anything bad these days that is not being tied to global warming.

Let's agree that the global temperature has been rising for most of the past century. (Most of it happened early in the century, and there's that little problem of a cooling from the 1950s to 1970s that throws a wrench into the man-causes-global-warming theory, but we'll ignore that.)

What we're left with an issue that is tailor-made for today's every-day-a-new-panic news cycle.

Crazy

As we've seen, you can pretty much link global warming to everything. Too much rain, too little rain, places being too hot, places being too cold, extinction of species (all the good ones), proliferation of species (all the bad ones), people flying too much, people driving the wrong kinds of cars, inflation, rising terrorism and on and on.

All these very scary news items come ready-made with great visuals: footage of hurricanes, dying polar bears and giant crashing glaciers are a great substitute for serious thought on the issue. And when that fails, we now have computer animation.

Special effects are better than ever these days. We can see a giant ape ice skating in Central Park, and in the next theater we can see what it would be like if global tides rose six feet over the course of the next 30 seconds.

Wow. Has there ever been anything more evil than global warming?

Global warming advocates and news outlets also know that if you can get children involved, you've got yourself a winning issue and a great fundraising cause. Throw in the near religious fervor with which Nature is elevated to god-like status, and you have all the makings of a crazy climate change cult that will be around for quite some time.

Speaking of madness, in Part Three–"It's the End of the World as We Know It"–you'll see how easy it is to get caught up in the hoopla of global warming thrills and frills.

Last Great American Whales
Speak Out

So 'round and around and around we go. Where the world is headed, ♫ nobody knows. 4

Hamlet Says: Some famous humans claim that global warming isn't a problem. But, if it is, that will not make a whale's day. They say mammals should not spend time worrying about the end of the world. Humans have taken advantage of whales before (see Free Willy and the three-time daily shows at *SeaWorld*). And doesn't history repeat itself? How can we trust humans when they have used us for meat, perfume and underpaid tourist attractions? It sounds fishy to me.

Harmony Says: No doubt about it, man has been a whale's worst enemy (remember Captain Ahab and poor Moby) and best friend (I wish I had a nickel for every concerned citizen who helped a beached friend of mine back into the sea). These days, you don't know whom to trust. Whales are probably best off taking care of themselves. If people have taken advantage of whales in the past, chances are they will do it again. It's all so much to worry about. I think I'll go for a nice swim.

PART THREE

It's the End of the World as We Know It

CHAPTER 5

EARTH: A Whole Lotta Shakin' Goin' On

"A developer is someone who wants to build a house in the woods. An environmentalist is someone who already owns a house in the woods."
–Dennis Miller

Did you know nature lovers everywhere love to tap into the symbolism of the Earth as Mother? But, if you happen to be in direct contact with nature, say in the middle of a 7.1 monster earthquake rumbling through your neck of the woods, you may think the nature lovers only got it half right.

The poet Tennyson was closer when he described nature as "red in tooth and claw." Left to her own devices, Mother Nature is much more like a cruel nanny than a kindly Mrs. Doubtfire.

Cold as Ice

It's quickie quiz time...

Which of these is not like the others:
a) rising tides
b) falling tides
c) floods
d) droughts
e) earthquakes
f) Al Gore's political career

The correct answer is (e). While we can classify all of these as natural disasters, the only one of these things that hasn't been blamed on global warming is earthquakes. But even on this point, some scientists tend to disagree. They believe that glacial earthquakes in Greenland warn of global warming.

As a matter of fact, until now, earthquakes are one of the few things that nobody has yet been able to blame on man's interference in the environment. Earthquakes occur randomly and spontaneously in nature like natural flavors in Dr. Pepper or steroids in a pro athlete's locker. No one knows what causes them.

Like most climate forecasts, conservative scientists believe earthquake predictions rank somewhere between tabloid horoscopes and tarot cards in terms of accuracy. One of the reasons for the many earthquake prediction theories is the sheer number of earthquakes. The world averages over 1,400 moderate to major earthquakes per year.

Each one of these earthquakes provides a ready-made news story, numerous photo opportunities and a fantastic chance for "experts" to generate confusion about the difference between correlation and causation.

The Cat in the Cradle

How can a struggling academic break through all the noise and get some government grant money? The solution is to compare earthquake data to something that occurs naturally, such as whales beaching themselves on the shoreline. Next, our scientist should highlight a correlation, suggest that there is evidence the migratory patterns of whales can forecast earthquakes. Finally, he'll need to assert that such an important finding most certainly warrants closer (and better funded) study. (Amateur scientists may find it easier to tune into your cat–or monitor cats worldwide–for a heads up about upcoming quakes.)

If you can get some politicians to sign off on this research, more power to you. The problem here is the same as with global warming. Despite the belief of both Democrats and Republicans that there are unlimited funds available from individual taxpayers and evil corporations, spending money on frivolous research means there is less money to attack real problems and more immediate concerns.

A short study of earthquakes reveals another important feature of today's media-driven society: publicity counts. Many people would probably agree the likelihood of the Big One in California is relatively high. It's just a matter of time–in the next minute, day, or week Californians seem prone to experience some kind of tremor. But then again, the rest of the country can't believe that people live in big houses on cliffs and on top of fault lines.

But wait a minute. Wouldn't our earthquake resources be better spent in the Mississippi Valley? *Ha-ha-ha-ha-ha-ha,* you say. *You're kidding, right?*

Well, only kind of. The folks at the U.S. Geological Survey website have slapped a pretty high percentage on a Midwest earthquake stating, "...the probability of a moderate earthquake occurring in the New Madrid (Mississippi Valley area) seismic zone in the near future is high. Scientists estimate that the probability of a magnitude 6 to 7 earthquake occurring in this seismic zone within the next 50 years is higher than 90%" (http://quake.wr.usgs.gov/prepare/factsheets/NewMadrid/).

The USGS site goes on to say, "Earthquakes in the central or eastern United States affect much larger areas than earthquakes of similar magnitude in the western United States. For example, the San Francisco, California, earthquake of 1906 (magnitude 7.8) was felt 350 miles away in the middle of Nevada, whereas the New Madrid earthquake of December 1811 (magnitude 8.0) rang church bells in

Boston, Massachusetts, 1,000 miles away. Differences in geology east and west of the Rocky Mountains cause this strong contrast."

The following picture highlights the relative area of impact of two comparable earthquakes: the 1994 Northridge, California earthquake and the 1895 Charleston, Missouri earthquake.

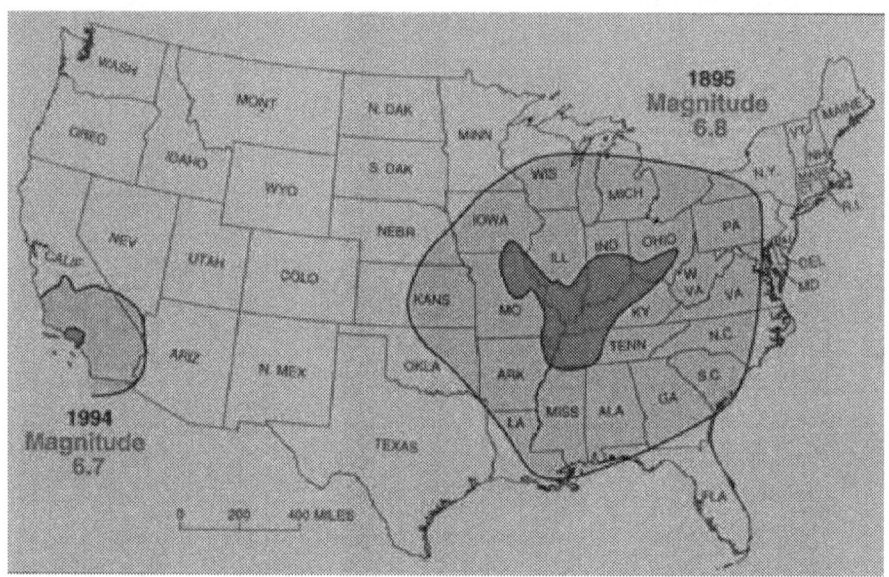

All Shook Up

Hey, did you happen to notice that the probabilities for these earthquakes occurring are significantly higher than any percentages attached to extreme global warming? With these kinds of probabilities attached and a hardy scientific consensus in hand, shouldn't we be convening a conference of politicians and publicity-seekers for a Kyoto Earthquake Summit?

There are important lessons to be learned from earthquakes regarding the global warming frenzy. More than three thousand people are estimated to have lost their lives in the Great San Francisco Earthquake of 1906. Keep in mind, the most severe earthquake to hit the San Francisco area since 1906 was the 1989 earthquake that forced cancellation of the third game of the World Series. The death toll as a result of that quake was about sixty people.

There is only one conclusion we can draw from this: God loves baseball fans more than gold prospectors. Or, perhaps, our salvation in the face of natural disasters, whether we're talking about earthquakes, global warming or used car commercials, lies in continued progress and improved technology.

Our conceptual faculty and our ability to adapt to changing conditions define our unique status in nature as "the rational animal." We would do well to play to our strengths and not fall prey to the fears of those with a clear agenda, whether they are former next-presidents or a new wave of religious earthquake predictors.

While East Coast folks can't understand how West Coast people live on fault lines, hurricanes are not in the vocabulary of Left Coasters. Well, that may change as the planet heats up. In the next chapter, you'll see the potential for new effects from winds making the rounds in unheard of cold and hot spots.

Last Great American Whales
Speak Out

I feel the earth move, under my feet. I feel the sky tumbling down.[5]

Hamlet Says: I can see it all clearly now: A trigger effect of hidden faults in Greenland creates a succession of earthquakes, which lead up to a potential great quake that hits the Pacific Northwest while we whales are migrating to Alaska in the summer. During the swarm of quakes, whales of all sizes, shapes and colors will be disoriented, strand themselves on the beach only to be swallowed up like a Milk Dud at the movies. A 12.5 superquake is imminent! What if we whales (what's left of us) unite and lead a climate crusade? Is it too late for a great whale march to save the Earth from its demise?

Harmony Says: Earthquakes aren't the result of global warming. This is nonsense. Earth changes impact both tremors and global warming. A key change is that the magnetic North Pole is migrating and will reside in Siberia in the next 50 years, according to some geologists and geophysicists. This may be why we whales are beaching ourselves more than before. I gotta get my fins on one of those new global satellite positioning systems. How the changes at the North Pole impact the magnetic force field, jet stream as well as plate movement are still debatable, say some humans who study quakes. So, if we ban stranding everything will look normal, right?

CHAPTER 6

WIND: Here Comes the Story of Hurricanes

"A lot of people think global warming is causing these terrible hurricanes. See, I think to stop global warming we should move in the other direction. We should move towards a second Ice Age. Follow me, if the glaciers are coming towards us at like an inch a year, then the government would have time to respond."
–Jay Leno

A consistent howling wind and sudden surges of hot air in an otherwise placid climate can mean only one thing. Climate change scholars are talking again. When it comes to hurricanes, the scientific connection to global warming is even more of a stretch than usual.

Most respected climate scientists attribute hurricane activity to a multi-decade cycle that has been in place for hundreds of years. Hurricane activity last peaked in the 1940s and began a decline that lasted into the 1970s. Since then (the last 35 years that global warming advocates conveniently cite as the time period for global warming-induced increases in hurricane activity), hurricane frequency has increased but only marginally according to statistics. Claims of increases in the intensity of hurricanes also appear, dare we say it, overblown. (Some reports suggest Hurricane Katrina may be an exception to the rule, or it could be that the only thing we're in danger of being swept away by is the global-warming-equals-more-hurricanes theory.)

Raindrops Keep Fallin' on My Head

Meanwhile, global warming's chief crusader, Al Gore, admits, "Yes, it is true that no single hurricane can be blamed on global warming. Hurricanes have come for a long time and will continue to come in the future. Yes, it is true that the science does not definitively tell us that global warming increases the frequency of hurricanes–because yes, it is true there is a multi-decadal cycle, 20 to 40 years, that profoundly affects the number of hurricanes that come in any single hurricane season."

One of the great things about being an advocate for action on climate change is that any change in the weather is a reason to leap up on to your hand-made, sustainable soap box and attempt to connect some phenomenon to global warming.

Has the Earth's temperature gotten warmer over the last 35 years from a cyclical low? Possibly, depending on how you measure it. Have hurricanes increased in frequency and intensity from a cyclical low over a similar time period? Marginally. Does that prove that global warming is causing it all? Hardly.

Current day Nostradamuses (or is it Nostradami?) are quick to cite a one-year spike in hurricane activity in 2005 as evidence of the cataclysmic future that awaits us if we don't do something about global warming right now.

Still, hurricanes make news and news draws groups with an agenda like Hollywood stars attract paparazzi. So we will most likely continue to see exaggerated claims of a connection between global warming and hurricanes. It sells like fad diets and health cure-alls.

Speaking of ailments, our theory on hurricanes is a bit different. When you get right down to it, aren't hurricanes really just Mother Nature going through PMS? It's probably worse because hurricane season comes only once a year.

Unpredictable behavior and severe water retention are only a few of the symptoms that let you know it's "that time of year" for Mother Nature. Talk about mood swings...one minute Mother Nature's throwing around trees and street signs. Then, she gets all calm and quiet during the eye of the hurricane. Next thing you know, she's freaking out again and tearing the roof off the Superdome. However, husbands, boyfriends and bureaucrats at government natural disaster agencies will tell you the most effective strategy for dealing with either hormonal or meteorological imbalances is to evaluate the immediate area and drive as far away as possible until it's over.

To combat the cyclical increase in hurricane intensity and frequency, our theory (as yet untested) is that we should shut off the Weather Channel altogether and switch to a 24-hour Lifetime network with a little bit of John Mayer music tossed in. In the end, we all need to be a little bit more sensitive to Mother Nature when she's bright orange on the Doppler radar. Whatever you do, try not to bring up touchy subjects like "global warming" or that whole "tsunami" episode from a few years back.

Realize that there's only one safe place to be during hurricane season. Hershey, Pennsylvania. Because that's where the chocolate is.

While we know gusty hurricanes, whether they're named after men or women, can wreak havoc around the globe, fire and global warming aren't something that humans (or whales) can ignore as you'll discover in the next chapter.

Last Great American Whales
Speak Out

Wipin' out... wipe out, ♫
Hey watch out, here we go.[5]

Hamlet Says: We whales are aware of our environment and what natural threats affect us. Heavy winds can wreak havoc on our lives because it can make the blue ocean noisy and this can affect our ability to sing our whale songs. Worse, baby whales can't hold their breath as long as us adults, so multiple trips to the surface can be scary and unsafe. And that's not all....

Some scientists say horrific hurricanes, fueled by warmer ocean temperatures, are the "smoking gun" of global warming. Loud hurricanes can increase shifts in weather patterns. What if it gets worse? What if we have to hear a category 8 event? Maybe if we whales all migrate somewhere safe like to *SeaWorld*, everything will work itself out.

Harmony Says: We whales can also swim away from bad weather while humans have to deal with the wind, water, and tornadoes in weird regions like sunny California. We are built to live in the water and can hold our breath for long periods of time, making it easier to survive a global warming hurricane. Super swells are a super blast. Whales can breach for joy while humans figure out this mess.

Rather than worry about a superhurricane, I'm thinking killer diseases. The word is that a hotter Earth may give rise to diseases, thanks to mosquitoes and rodents. Gee, I wonder if whales are affected by these potential human health risks linked to climate change? Maybe while we fight Rodent Fever or Gigantic Mosquito Plague, scientists will forget about global warming and save mankind.

CHAPTER 7

FIRE: I've Seen Fire and I've Seen Rain

"Intellectuals cannot operate at room temperature."
– Eric Hoffer

If you've ever tried to start a fire for a barbecue, you understand that conditions have to be just right in order to get the charcoal going. So you know global warming must be an awesomely powerful force since climate change advocates are proposing a direct link between global warming and an increase in wildfires throughout the world.

The fact that the weather can cause endless tracts of land to spontaneously combust while you struggle to get charcoal to catch fire with a butane torch and a three bottles of lighter fluid tells you all you need to know about the power of nature and the evil effects of global warming.

Come to think of it, what are you doing having a barbecue with your family anyway, you insensitive planet-killer?

Don't you realize burning charcoal for a typical barbecue releases .000000000002$^{(-10)}$ milli-somethings of CO_2 into the atmosphere? Stop and think what could happen if everyone in the country had summertime barbecue with their family. The global temperature might possibly go up .0000006 degrees centigrade at some point over the next 60 years. Those barbecued ribs don't taste nearly as good when you realize the danger you're subjecting your children's children to, now do they?

Well, actually they do. We've found that living on the razor's edge of climate change danger merely enhances the tangy essence of tomatoes and serves to complement the smoky spiciness of chipotle. (See Hamlet and Harmony's recipe for *Global Warming Barbecue Sauce* at the end of this chapter.)

Light My Fire

If having to deal with your neighbor's screaming kids and listening to advice from your father about when to turn the chicken isn't enough to deter you from damaging the ozone, perhaps a list of equivalent planet-saving choices will make you switch to cereal at this year's family picnic. But then again, maybe not.

Consider that, for the same amount of greenhouse gases you are sending into the troposphere with a typical barbecue, you could:

–Hold screenings of *An Inconvenient Truth* without popcorn, increasing your environmental awareness and at the same time saving valuable corn for use in producing ethanol
–Fuel three stretch limousines to take politicians to hearings about global warming (if they're seriously concerned about global warming they'll limo pool)
–Televise 21 minutes of those same hearings or compile 3122 ill-informed sound bites for use on evening news shows
–Fuel one private jet ride for actors, musicians and other celebrities to appear at various photo ops telling you we are "at a tipping point and must take action on climate change *right now!*"
–Produce 3802 recyclable clipboards for use by students collecting signatures on petitions urging "someone" to "do something" about global warming

Put Out the Fire

Like many of the effects attributed to global warming, fires have been occurring for many years–with and without man's help. Yes, surprisingly, the world has had its share of fires before you were born and before the industrialization of the 20th century.

American Indians, long held up as the models of sustainable living, regularly burned forests in an effort to create a more hospitable environment for the plants and animals upon which they depended.

In nature, lightning causes a number of fires every year. Amazingly, no one has yet made the claim that global warming has increased the amount of lightning. The upside is, it's not too late to form your own environmental organization and get your fair share of government grant money. The downside? Whether we like it or not, fires actually serve a purpose in the environment. A forest not gardened out or not subject to brush-clearing fire on a regular basis will develop a ground cover that can cause an extremely hot, low fire that sterilizes the soil when a fire eventually does occur.

No one is suggesting that we let fires burn out of control, but there are certainly other alternatives to forest management than bringing industrial progress to a screeching halt.

Into the Fire

The main complaint in the thinly stretched global-warming-leads-to-more-fires chain is that fires created by global warming will have a negative effect on biodiversity (the total count of species and ecosystems in a given location).

Biodiversity is one of those fancy six-syllable words which sounds like it's good for you but is hard to put into practice...like getting a root canal or developing an appreciation for opera.

On its surface, the idea that we can somehow mastermind the whole assortment of plants and animals on the planet is absurd. Attempting to regulate the percentage of species in relationship to each other is a tad arrogant. We can't control the combination of plants and animals in our yards, and yet climate change doomsayers want to try to manage it on a global scale. *Memo to climate activists:* When you figure out what the precise mix of mosquitoes, ants, worms, birds, flowers, weeds, squirrels, trees, dogs, cats, bacteria and humans should be on your own lawn, give a save-the-planet organization a ring.

Still, there are no shortages of volunteers who are willing to step up to accept vast amounts of money to try to balance human needs with those of 80,000 species of insects and other animals. Whether it's Congress (motto: "Relax, there's enough pork for everyone") or the United Nations (motto: "Why doesn't anyone listen to us?"), everyone wants a crack at playing God.

As humans, we've done a pretty good job at taking care of ourselves and the occasional house pet. Let the rest of the animals take care of themselves. It's what Darwinism is all about. Trust us; even if they *could* talk, lions, sharks and other predators would tell you they don't give a rat's (or a gazelle's or a tuna's) tail about preservation of any species other than their own. Maybe they know something.

As promised, here is the recipe for Global Warming Barbecue sauce. It goes great with krill (available at your local health food store) and other endangered species.

Hamlet and Harmony's Global Warming Barbecue Sauce

¾ cup of apple cider vinegar
4 tablespoons sugar
1½ teaspoons tomato paste
4 tablespoons ketchup
3 tablespoons molasses
2 tablespoons minced garlic
1 - 2 tablespoons chopped canned chipotles in adobo
1 tablespoon minced peeled fresh ginger
1 teaspoon salt
Mix sauce ingredients in a 1 quart or larger saucepan and simmer rapidly, uncovered, stirring occasionally, until sauce thickens and is reduced to about 1 cup. Use immediately or store frozen up to six weeks or the until the release of the next global warming issue of *Newsweek.*

So the question remains, will global warming cause a Noah's Ark type of flooding? In the following chapter, you'll find out everything that you were afraid to ask about rising tides of seawater and hysteria.

Last Great American Whales
Speak Out

> Woo ah, mercy mercy me, Things ain't what they used to be.[7] ♪

Hamlet Says: Volcanoes release huge amounts of water and carbon dioxide. Too many eruptions of supervolcanoes can increase the CO_2 levels in the air. It dissolves in our ocean and can maybe dissolve whales, too. This is all scary, sad and depressing. It's enough to make a whale have a nervous breakdown. My world has become "weary, stale, flat and unprofitable."[1] Kind of like the United Nations.

[1]*Hamlet*, Shakespeare, Act I, Sc. II

Harmony Says: Almost two hundred years ago, after the eruption of Tambora, temperatures dropped, causing crops to die and famines in America and Europe. So, yes, volcanoes can change the climate of the earth. But while eruptions are beautiful like a breaching whale, I think I may fall victim of seasonal blues year-round if we end up having to deal with the build-up of pollution.

CHAPTER 8

WATER: Take Me to the River

"Yesterday, a group of scientists warned that because of global warming, sea levels will rise so much that parts of New Jersey will be under water. The bad news? Parts of New Jersey won't be under water."
– Conan O'Brien

When it comes to global warming and water, the big question on everyone's mind is, "Does global warming mean there will be more floods or more droughts?" If you've followed the science so far, the answer should be clear. "Yes." It's the planetary version of a Catch-22.

In this sense, global warming is like being a teenager–you can't do anything right. There is a rapidly building scientific consensus that anything bad that happens anywhere in the world can be traced directly back to either global warming or rebellious teenagers.

Blame It on the Bossa Nova

Not only can teens use global warming as a ready-made scapegoat, so can parents, grandparents, and just about anybody. Yes, the family dog now has been given a reprieve thanks to global warming.

Note to teens: Global warming is the best thing to happen to you since text messaging. Most stupid questions your parents ask you now can be easily answered with two words: global warming.

Examine the following results of a survey compiled by leading experts in adolescent behavior and see how easily "global warming" answers every question.

Top 10 Stupid Questions Parents Ask Their Kids:

10. Why can't you be more like your sister?

9. How could you possibly go over 500 minutes a month on your cell phone?

8. Why can't you get better grades in school?

7. Why does a 16-year-old need birth control pills?

6. Where did this pot come from?

5. How did this dent in the car get here?

4. What were you kids doing downstairs till 3:30 in the morning?

3. Why are we getting all this spam from X-rated sites on the computer?

2. What have you been doing in the bathroom for the past half hour?

1. What made you think that taking your father's rifle and shooting up fourth-period gym class would solve your problems?

For variety, feel free to substitute "climate change" or the slightly more cumbersome phrase "planetary temperature increase." These words become even more effective when complemented by traditional actions such as surly, disinterested shrugs, mumbling and failure to make eye contact at any point during a half-hour conversation.

Beyond the Sea

When it comes to water, there is no shortage of disasters that can be blamed on global warming. One of the interesting things about nature is that it has a way of extracting its toll on those who try to deny its realities. You can build a city below sea level, but you'd better make sure you don't skimp on the levees. You can construct a 6500-square foot mansion on a muddy hill overlooking the Pacific, but you might want to think about what happens when there's a rainstorm. And, let's face it, if you live in a small village in India that is repeatedly washed out by flash floods...your second home should be a raft.

Our particular advantage as humans is our ability to mitigate the effects of nature. We can pump water to dry areas, we can cool down buildings in hot areas, and we can dress up poodles in formal wear and train them to walk around in the circus. With that kind of ingenuity, we can probably figure out how to adjust if the temperature goes up a couple of degrees over the next 50 years.

Surf City

In the meantime, we can also create all kinds of special-effects videos showing water covering half of California or lame and adorable polar bears being stranded at sea. It's all very scary stuff. It's just that when you stop to think about it, none of it is very likely to happen.

Think of all the artificial crises that we've faced over the past generation: Y2K (the world was going to come to a halt and there would be mass riots); road rage (the nation's highways were going to become a shooting gallery); and the West Coast slipping into the Pacific Ocean (one reason why Las Vegas is the fastest-growing city in the country–it could become oceanfront property at any moment).

Have any of these events turned out to be anywhere near as dire as the media hype at the time? Of course not. Headlines like "Y2K is No Big Deal" and "Shrinking Lakes Due to Natural Evaporation" don't sell papers.

The lesson to be learned, going back to the ancient prophecies of Nostradamus, is that the one shortage we'll never face is a shortage of people trying to convince us that the world is coming to an end.

And last, there is talk about rising sea levels flooding coastal towns and major cities, such as Los Angeles and New York. We even have people worrying that the next step after that will be for it the water to freeze and bring us right smack into the next Ice Age. In the next chapter, find out if it's time to wax your skis.

Last Great American Whales
Speak Out

Raindrops keep fallin' on my head, they keep fallin'...⁸ ♫

Hamlet Says: Some humans say sea levels are rising fast each year. Blame it on the expansion of the top layer of our oceans as they warm and the melting of the polar ice caps. By 2050, this may cause flooding in coastal regions, and places like London will be in trouble, too. While I don't spend time in England, what about our fellow whales? This is all too much for me to handle. What if I live long enough and have to die in a global warming disaster? Maybe I'll just leave it up to humans to figure it out. After all, I'm just a whale.

Harmony Says: Personally, I'm more prone to linger in Alaska and Hawaii. Flooding may cause problems for humans. Gosh, if there is massive coastal flooding or if the sea consumes cities like San Francisco–we whales will be even more disoriented–as if San Francisco isn't hard enough to find your way around already! We will end up beaching ourselves on a no-name beach or Alcatraz or Chinatown. It's enough to make a carefree whale never want to reproduce, migrate, or live to 90. I feel a blues song in the making.

CHAPTER 9

ICE: Ice, Ice, Baby

"You never know who your friends are until the ice breaks."
—Eskimo proverb

When it comes to scare tactics, global warming advocates will travel to the ends of the Earth in an attempt to bolster their arguments. Case in point: all the excitement over ice shelves collapsing in Antarctica.

Seriously, do you even care what goes on in Antarctica? With everything else going on in the world, do we really need to worry about losing some ice? The only time running out of ice is a problem is when it's 2:00AM on a Friday night and the party is just kicking into gear.

Antarctica is the last great underdeveloped area on our planet. Let's stop being so negative and start thinking of all the good things that would result if things were just a little more toasty down around the South Pole.

It's Cold as Ice

Let's face it, most people hate winter. Oh, sure, it's nice to have the seasons change and to have snow around Christmas, but you'd get a lot more tourists to visit Antarctica if there were girls in bikinis strolling along warm beaches instead of six bearded guys with weather balloons and a 95-day supply of canned meatloaf.

We'll grant you penguins are very cute, and they do add a certain air of formality to the discussion about global warming, but, if they could talk, don't you get the impression they wouldn't mind if it things were just a *bit* warmer?

Despite dire warnings to the contrary, the fact of the matter is that the trend in Antarctica has been toward *cooler* temperatures over the past 50 years. On glaciers other than those few selected for photo-ops, the ice is actually thickening.

You don't hear too much about that because it's not as exciting as watching 40 tons of ice collapse into the sea and it really messes up the generally accepted global warming theory that the greatest temperature increases should occur at the Poles. Ooops.

Talk of collapsing ice shelves quickly leads to a discussion about a rise in sea levels around the world. If all this ice melted into the sea, wouldn't it mean that Manhattan would be underwater in about 45 minutes from now?

Well, not really. For the same reasons that your ice-filled glass of *Mountain Dew* doesn't overflow when it's left out in the sun. When ice melts, the water added from melting is exactly equal to the amount of water that was displaced by the ice when it was floating.

It's the same principle with an ice shelf floating in the ocean. Even if it melted, sea levels wouldn't increase because the water from the melting shelf would merely be replacing the water that had been displaced when it was ice. So let's all just chill out about the dangers of melting ice shelves.

Hot Fun in the Summer

Concern about retreating glaciers is one of the wackier elements of the whole global warming discussion. The Earth had its last big Ice Age about 10,000 years ago. Since then the planet has been in an interglacial period in what has been a regular 10,000-12,000 year cycle of cooling and warming. We're no scientists, but it seems to us that when we're coming out of an Ice Age, it's reasonable to expect glaciers to be retreating.

You want to talk scary? What about this tendency for temperatures to increase as kids get closer to the end of the school year and adults begin planning their summer vacations? It is absolutely irresponsible that we expose the human race to this kind of warming...just at the time when our children are spending more time playing outside and adults are trying to get in shape for beach season. The time to act is now. It's a moral imperative. Please join our cause and help us stop summer before it's too late.

All kidding aside, even if you believe global warming is a concern for humans and animals worldwide, living each day repeating the mantra, "Be worried," or marching to save the planet isn't going to make it all go away. If you want to do your eco-friendly part, be our guest–especially if you feel it's better to be doing something rather than nothing. Some scientists will even tell you that having a purpose in life boosts longevity.

Our best advice: live life for today because a monster asteroid could hit the day after tomorrow. And then, you'd be in heaven saying, "I should have had more fun in the summer before the world ended."

Last Great American Whales
Speak Out

There's something happening here,
What it is ain't exactly clear.

Hamlet Says: People who study earth's climate have found countless glacial earthquakes caused by quick movement of Manhattan-sized blocks of ice in Greenland. Worse, warmer temperatures caused by global warming could melt the Arctic and Antarctic sheets—our favorite hangouts—sooner rather than later. If this global warming worst-case scenario is in the cards, I believe it's time for all good whales to prepare themselves. It's time to reread Emerson's *Self-Reliance*.

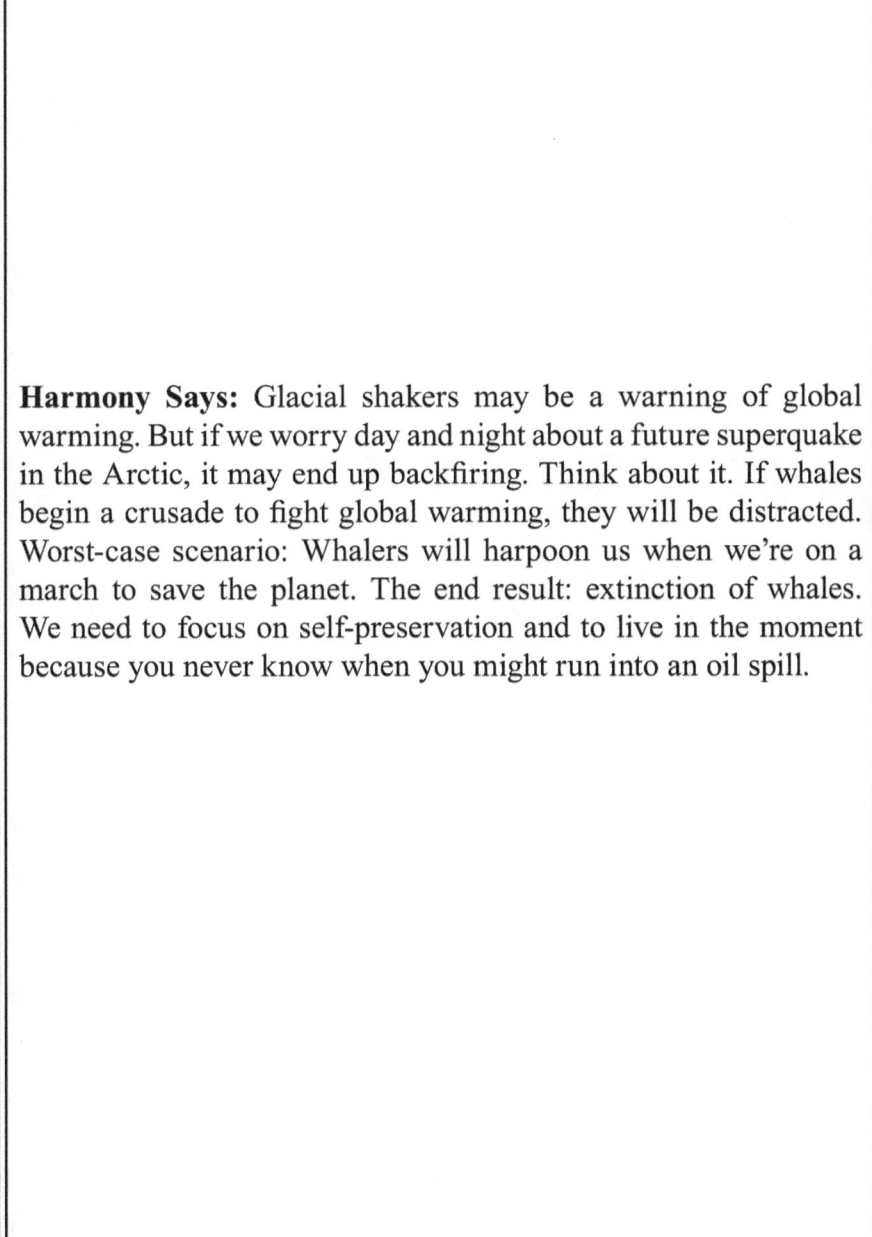

Harmony Says: Glacial shakers may be a warning of global warming. But if we worry day and night about a future superquake in the Arctic, it may end up backfiring. Think about it. If whales begin a crusade to fight global warming, they will be distracted. Worst-case scenario: Whalers will harpoon us when we're on a march to save the planet. The end result: extinction of whales. We need to focus on self-preservation and to live in the moment because you never know when you might run into an oil spill.

PART FOUR

I Will Survive

CHAPTER 10

Who'll Take the Dogs Out?

"I've seen a heap of trouble in my life, and most of it never came to pass."
– Mark Twain

So, what can *you* do to stop global warming? Not a whole lot. Let's troubleshoot this worldwide dilemma for just a second. We're talking about the weather and we're talking about the entire Earth.

But what if we *all* did a little something?

Sadly, your chance of meaningfully impacting global climate is about the same as changing the Earth's orbit in the event a random asteroid was heading our way.

Great Balls of Fire

Take a hypothetical look at how the asteroid scenario could become the next global crisis....

Somewhere in a lone observatory atop a mountain, a solitary scientist peers into a giant telescope. He notices a small speck moving across his field of vision somewhere in the vicinity of Jupiter. The first strains of scary music start to play in the background. The scientist zooms in on the speck and…it's an asteroid! And it's headed our way!

Your children and your children's children might be at risk. After all, didn't a giant asteroid wipe out the dinosaurs? What if it happened again? What can we do? Is there no hope for the planet? Are we doomed as a species? Do we not have a moral imperative to try to save the planet for our children?

At this point, you can be pretty sure some politician (citing a rapidly growing scientific consensus) would suggest that we all fly to Australia and jump up and down to help alter the planet's path around the sun.

And, if you're really all that concerned, you should feel free to book a ticket. But, please, let the rest of us to worry about more important things like what kind of white wine goes with that new salmon recipe we just read in *Parade Magazine.*

This is not to say that there aren't some things you can do to delude yourself into thinking that you can make an impact on global warming.

Wouldn't It Be Nice

Just think about it. What if everyone took some steps to stop global warming? At the very least, you might make a few friends, and you'd get the chance to pester your elected officials. Meanwhile, if you're busy dealing with day-to-day woes, we've got you covered. Read on–discover 12 things you can do to *feel* like you're doing something about global warming. Life is a beach, isn't it?

1) Get your own cow. Today's modern dairies are responsible for [insert scary, unsubstantiated percentage here] of the greenhouse gases in our atmosphere. Pasteurization, likewise, is overrated. In the pastoral times of the Middle Ages, when the air was clean and windmills and babbling brooks provided most of the power, no one had pasteurized milk. Sure, there was rampant dysentery and disease, and the average life span was one-third what it is today, but that's a small price to pay to make sure we leave a viable planet for our children.

2) Stop all activity when it gets dark. While owning a cow is a start, turning off the lights is a must-do. Most of the power that's used in this country is used after the sun sets. Televisions, stereos, and vibrating hotel beds all help to make America the most energy intensive country in the world. We need to start now, right now, if we're going to try to impact the planet's temperature somewhere beyond the 23rd decimal place. The Amish have lived this way for centuries. And if they have survived living the simple life, we can vow to die living a boring life, too.

3) Walk to work. Early to bed and up at dawn to milk the cow before you face your new, improved four-hour commute to work in the morning. Only the most selfish person would suggest this isn't a reasonable price to pay for helping reduce CO2 emissions back to the level they were at the time of the Pharaohs.

4) Eat all your food raw. Getting back to basics is healthy and, yes, this gets sticky if you include meat and poultry. *Note:* There is a small risk of catching mad cow disease, but people are already going nuts over a two-degree rise in temperature over the next hundred years, so how much worse could it be? As it turns out, the symptoms of mad cow disease in humans already pretty closely mimic the effects of being a climate change activist, including (but not limited to) "memory loss, emotional instability, including inappropriate outbursts, and severe rapidly progressive dementia."

5) Use stone tablets instead of Post-It notes. If you can do raw food, you'll have a little more time to express your thoughts. But instead of sending an e-mail or writing on a Post-It note, go back to the Stone Age and do what cavemen did–use rocks. You'll feel connected to nature.

6) Communicate by carrier pigeon. We know it seems like you couldn't live without e-mail or your cell phone, but computers and cell phones are among the most energy intensive things to recycle. And remember, we're trying to eliminate using any energy so that…well, we just are. Just play along, okay? If you are allergic to birds, as an alternative, you might also consider hiring a messenger boy to replace your current long-distance carrier.

7) Unplug all the appliances in your house before you go to work…Before you begin that morning commute and, after you've answered all the messages left by carrier pigeon, unplug your appliances. They use electricity even when they're not turned on. Your annual electric bill could be tens of cents lower and you could single-handedly save enough energy to power your wristwatch for 48 months.

8) … And unplug all the appliances in your house before you go to sleep. While you're at it, unplug all the appliances in your neighbor's house and at the houses of complete strangers. Trust us, rock salt blasted at you from a shotgun at close range only stings for seven or eight hours. It's a small price to pay for taking the moral high ground on the issue of global warming.

9) Opt out of your hospital-based HMO and find a good local witch doctor. Staying healthy and saving the planet take creativity and moxie. Modern medical machines like MRIs and CAT scans contribute to America's energy addiction and the rising cost of healthcare. By patronizing your local witch doctor, you'll be doing your part to solve two of the country's most pressing problems. *Tip:* Ask for the generic leeches. They're every bit as effective in bloodletting as the name-brand leeches.

10) Plant a tree in every room in your house or apartment. Good-for-you trees not only help to reduce your carbon footprint (making you like an invisible Ninja to Mother Nature's sensitive ecosystem), but they also provide shade from those over-priced yet incredibly brilliant energy-efficient bulbs you're no doubt using throughout your house. Don't forget to insulate heavily around the hole in your roof where the tree sticks out.

11) Clean yourself with rubbing alcohol. Don't despair! Saving energy doesn't mean hygiene has to take a holiday. Showers account for two-thirds of all water heating costs. If you're serious about saving the planet so that your children can screw it up without any kind of a running start from your generation, eliminate showers altogether and loofah with rubbing alcohol or Purel waterless hand cleaner. The scent of alcohol will let people know that you are doing your part to save the planet.

12) Set your thermostat two degrees higher in the summer to cut down on cooling costs. Last, but not least, setting your thermostat two degrees higher will illustrate (in just a few short hours, right in your own home) the catastrophic temperature change that is predicted to happen over the next hundred years. This should go a long way toward easing your mind about climate change Armageddon and free you up to worry about more important things like remembering to milk the cow.

Last Great American Whales
Speak Out

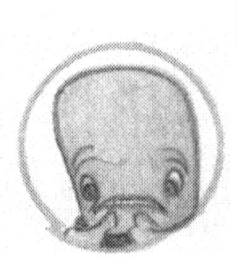

To be or not to be:
That is the question.[10]

Hamlet & Harmony Say:

Here is a must-have checklist for a Global Warming Watch. While we have experienced the five stages of global warming grief: denial, anger, bargaining, depression and acceptance–humans can do the same. It's called "survival instinct."

Rather than keep your head in the sand like a beached whale, do as we plan on doing–go with the flow on land and off. In whale talk that means–life goes on.

And note, global warming can be exciting and thrilling, like a great whale watch. To have a fun, safe view of the Earth's changes, you have to stay cool, calm, and not have preconceived expectations. Nobody knows for sure when or if the world is going to end due to the planet getting hotter.

- Do understand that you're not going to see doomsday in action! While global warming tours often can show real-life effects of storms, droughts, and floods, you may see one of these disasters or none at all.

- Bring along snacks and beverages (hot and cold) to make your tour more comfortable. You never know what you're going to get, a cooling or warming episode.

- Speaking of temperature changes, wear layers of clothes so you will be comfortable in the sunshine, rain, snow or sleet.

- You can snap photos, but don't expect to see special effects like your favorite disaster movie. It's best to take a camcorder and have an endless vacation to guarantee the best results of a changing planet.

- While people often claim to see the effects of global warming, we don't "guarantee sightings." But we do suggest before you leave this planet, come back again to see if the sky is falling.

Afterword

Give It Your Best Shot

Want to walk the nature walk? Whether your mission is to stop global warming and save the planet (or whales), here are a few hot tips to make a personal statement in the wide world of humans and mammals.

#1: Hybrid, That's All We Ever Hear: Why drive a hybrid car when what you should be driving is a rickshaw? Talk about status. Not only will you be just as trendy as a Hollywood starlet by adopting kids from a third-world county, you'll also be helping to reduce our dependence on foreign oil. **Harmony & Hamlet Tip:** It will keep our oceans less polluted, more full of krill.

#2: Expect More, Pay Less: Don't be fooled by big corporations who are posturing for your green dollar. Rather than buy pricey clothes, forego them altogether whenever possible since the temperature is rising. **Harmony & Hamlet Tip:** Whales go all natural and therefore are not adding to the man-made global warming crisis.

#3: Forego Lawn Mowers: Want a totally reasonable solution to maintaining your lawn? Try goats. Companies like Goats-R-Us (www.goatsrus.com) charge only around $700 an acre for their service. While that may sound expensive, that cost includes transportation, supplements, healthcare and insurance for the goats. You'll feel better knowing that you're helping the environment. On the other hand, you'll probably feel worse that the goats have a better benefits program than you do. **Harmony & Hamlet Tip:** Take up whale watching and see how whales survive up to 80 years without manicured yards or perks to make their day.

#4: Travel S-l-o-w-l-y: Eco-tourism is the latest fad in the travel industry. But wait, you weren't really thinking about flying or driving your SUV down to the Everglades, were you? The only natural way to head to Florida on your spring break is to walk. Assuming you walk at a reasonable pace and don't dawdle at rest stops, you should be able to make the trip in two weeks without sleeping. Not close enough to nature? Refer to #2 Whale Tip: Dress down for the trip. **Harmony & Hamlet Tip:** Whales sleep eight hours per day and travel thousands of miles without using air-polluting fuel and creating havoc behind them.

#5: Pay More, Eat Bland: Whole foods and other organic supermarkets specialize in over-priced, tasteless food. But that's a small price to pay for doing good things for Mother Earth and striking a blow against big evil corporations. Oh, did we forget to mention that 95 percent of all organic fruits and vegetables come from three massive eco-friendly farms in California? Way to stick it to The Man, Eco-Friend. **Harmony & Hamlet Tip:** When facing global warming danger–do as whales do: ration krill and pray there is enough to go around without having to harpoon your fellow man.

As an added bonus, we thought we'd share with you a few of our favorite websites on the topic of global warming.

For more information, Hamlet recommends:
www.climatecrisis.org   www.stopglobalwarming.org
www.savethewhales.org   www.greenpeace.org/international
www.sierraclub.org

For more information, Harmony recommends:
www.junkscience.com   www.sepp.org
www.greenspirit.com   www.heartland.org/index.cfm
www.climatescience.org.nz

WHALE NOTES

[1] *Hamlet*, Shakespeare, Act I, Sc. II
[2] "Here Comes the Sun," The Beatles
[3] *Hamlet,* Shakespeare, Act II, Sc. II
[4] "Ball of Confusion (That's what the World is Today)," The Temptations
[5] "I Feel the Earth Move," Carol King
[6] "Wipeout," Beach Boys
[7] "Mercy Mercy Me (The Ecology)," Marvin Gaye
[8] "Raindrops Keep Falling on My Head," B.J. Thomas
[9] "For What It's Worth," Buffalo Springfield
[10] *Hamlet*, Shakespeare, Act III, Sc. I

About the Authors

Cal Orey is a popular author-journalist. She holds a B.A. and an M.A. in English (Creative Writing) from San Francisco State University. Over the past twenty years, she has written hundreds of articles for national and international magazines, including *The Writer, Woman's World, The Writer*, and has been contributing editor for *Complete Woman*. She specializes in writing about health, relationships, pets, and science. Ms. Orey is the author of several books, including *202 Pets Peeves: Cats and Dogs Speak Out on Pesky Human Behavior*, and *The Man Who Predicts Earthquakes: Jim Berkland, Maverick Geologist–How His Warnings Can Save Lives*. A native of the San Francisco Bay Area, she lives in Northern California. Her website is www.calorey.com.

Mark Jabo is a writer, comic and financial consultant who spends time in Florida, New York and Maryland. Originally from Philadelphia, Mark has lived and worked all over the world including Tokyo, London and Australia. He was a founding member of *The Comics Roundtable*, a writing and performance group headed by Emmy-award winning comic, Paul Wagner. He is a contributing writer for *TheCheers.org* webzine and was a finalist in Humor Press "America's Funniest Humor" contest. His short stories have been featured on Apollo's Lyre and Long Story Short websites–both sites have been honored by *Writer's Digest* as among the "101 Best Writing Sites." Mark's climate change blog is: www.getmehot.blogspot.com

www.ingramcontent.com/pod-product-compliance
Lightning Source LLC
Chambersburg PA
CBHW022105170526
45157CB00004B/1490